The Religion of One
(The Pighne Theory)

By Marcas Major

Published by MP2ME Enterprise
For information, please contact:

MP2ME Enterprise
16754 SE 45th Street
Issaquah, WA 98027
http://www.mp2me.com

ISBN: 0-9717947-9-0

Printed in the United States of America

Keep asking, "Why?"

Table of Contents

The Religion of One

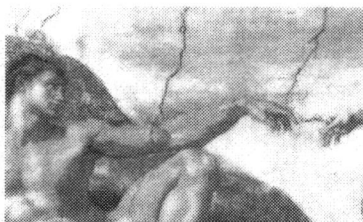

By Marcas Major

Introduction

"String Theory" tries to describe the smallest particles in the universe. The strings represent squiggly, stretchable strands of existence. They are proposed to be the building blocks for subatomic particles— that make up quarks; that create the components of an atom; that make up molecules; that make up life.

There are problems with the string theory physicists have trouble explaining. Five string theories are mathematically sound. Introducing space and time into a formula where space and time never existed, in the beginning, could lead to variations in formulas that try to describe the essence of existence. These different formulas become interpretations of fragments of something that was once whole. These formulas do not necessarily represent the object when it was whole but only represent the fragments of the entities after the "Big Bang". The "Big Bang" theory has been proposed to be the way our universe was originally created.

Analysis of such theories should be done by individuals other than physicists. A different point of view can help shed light on why we exist in the universe.

I came to the conclusion, thirty years ago, that people believe in God because they need to. This article is an attempt to describe why people need to believe in God from the view point of the "String Theory".

The Big Bang Theory

If you type in "religion of one" in quotes in a search window, on-line, you can find a common theme running through many of today's religions. The common theme is that there is only one God.

Let's take that idea one step further. God is the universe. That isn't a giant leap in faith based on common scientific principles regarding the Big Bang theory.

In the beginning there was no space and time. Dimensions did not exist. In the beginning there was just one entity that included everything, everywhere, always, and forever. It is easy to come to the conclusion this all inclusive entity could be the one and only God proposed in many religions. It follows that this all encompassing entity would also be "all knowing" since it included everything that ever was before it ever existed.

When the Big Bang occurred did the existence of God as one cease to exist? It is possible that the Big Bang created the universe and the universe is God. With the introduction of the three dimensions in space and time God could no longer be One. Maybe this was God's conscious choice. Maybe God's greatest sacrifice was to become the universe and share all that was seen and unseen before the Big Bang with all objects in the universe. My wife does not like this idea. She says that it destroys the idea of an entity that you can love and worship. I am only following a train of logical thought. I am not trying to destroy centuries of modern thinking. I also believe there is a consciousness that exists in a "fourth dimension" and may represent God's whole existence, as one, in another dimension than our own. In our universe the Big Bang theory proposes a large explosion as the creation of the universe. With any large explosion there is randomization of particles. If God was one and an explosion occurred then in our three dimensional world it would appear to me that God's particles would be spread across the universe in a random pattern. With every action (explosion) there is an opposite and equal reaction. This opposite reaction may have released God into a different dimension as the fully conscious entity we worship today.

Every subatomic strand in the universe can simply be thought of as a piece of the One— a piece of God with the awesome power and knowledge of millions of billions of stars packed into each stretchable, squiggly strand that the "String Theory" proposes. The smallest particles in the universe are the essence of God. Each object in the universe has these strands of God. Perhaps this means that each object in the universe has a soul.

It is possible that the smallest particles in the universe (the essence of God) have only one goal— to become One again. It may be what drives the existence of life. Any great effort in life is driven by the need to become one again. Life, in a religious sense, is all about getting back to being one with God.

The Need to Get Back to One

In my opinion the nuclear bomb becomes the essence of pure evil. The nuclear bomb is an attempt to recreate the Big Bang with the resulting wasteland of pieces of what was once whole. The efforts to become one with God are shattered into random pieces that take centuries to heal. Even after nature has been blown apart by nuclear forces a healing process does begin and nature marches on in the attempt to become one with God again.

The Existence of Life

We are still compelled by the essence of God in each of us to become one— to become complete. This could be God's greatest gift. The gift to every object in this three dimensional universe could be the ability to always strive to be whole and be one with God using the very essence of God that is within all of us.

It could be argued that nature's attempt to become one with God culminated in the creation of the DNA structure. These small strands are the building blocks of life. DNA is protected in each living thing with the sanctity of a holy object. The same basic structure for DNA exists in all living things. It is just subtle variations in the structure of DNA that results in the plethora of life on earth. With the essence of God defining our very existence we innately have great powers to move towards being one with God. Unfortunately we live in a universe with dimensions in time and space. These laws never existed before the Big Bang. Any attempt to get back to the one God is burdened by the laws of a three dimensional universe that never existed when God existed as One.

Wave Theory

It has been proposed that variations in oscillations of "strings" are what create the various elements in the universe. Different frequencies within each "string" define the great array of objects in this universe. A more in-depth study of waves can help us further add to our knowledge of the originations of the universe and life. The properties of a wave define the existence of our universe. Waves oscillating at different frequencies can create or destroy objects in the universe. Waves can be used to shatter glass, heat objects, create the fabric of life and destroy the fabric of life. How waves behave within variations of time and space may lead us closer to becoming one with God. Perhaps that is why it is believed we become closer to God when we die. When we die, time and space are eliminated. Only the essence of God is left in our bodies when we die. Some people have come back from a dead state and have claimed they saw a wave of light they knew was the path they were supposed to follow. They claimed this light would lead them to God.

Life may not exist because nature randomly combined molecules of matter together and accidentally formed life. A better explanation may be that there is a logical, selective process going on at a subatomic level. How do these tiny strands of God's existence after the Big Bang combine with each other in an attempt to become whole again? I believe the answer may be found by understanding how waves interact with each other at different frequencies and at different levels of time and space. Mathematically it becomes a search for the original formula that described the universe as One before the Big Bang. Wave research could be the path to defining the way God existed as One.

Conclusion

What I have just presented is a celebration of all religions and their attempts to become one with God. It seems fitting that I lay down these final words on the eve before Christmas. It is a time we reflect on the birth of Christ and his teachings. Could there be another meaning of Christ's death on the cross? What if the cross symbolizes our three dimensional universe? What if Christ was trying to show us that God's greatest gift was to die as One and be reborn as the universe? I have detailed how the power and conscience of One could be within us all. The smallest, most fundamental entity of our existence could simply be the essence of God.

The Roots of Our Existence

Introduction

Here is a great story and it happens to be true. The ending to this story just happened last night — 7/28/2005 at around 3:00 AM. I am a normal guy. I have a BS in Fisheries Management, an MBA in Finance, and MCSE and MCDBA Microsoft certifications. I take things I hear and try to come to logical conclusions about how they might have occurred.

Hope and the Paranormal

My sister (half-sister) visited her 86 year old dad in Vancouver, B.C a couple of weeks ago after finding out his colon cancer had come back. His doctors decided not to take any action on the colon cancer due to his age.

She is accustomed to bringing her dog (a Bichon Frise) every where she goes. It is her baby after having two kids. She has the classic minivan that is good for bringing kids to soccer, etc. Her dad decided he wanted to go out to eat at a restaurant near the Vancouver International Airport. This is not the best part of town. She parked the car in a parking lot near the restaurant and left her dog in the minivan while they went and had a nice lunch together.

When they came back to the minivan after lunch they found the window had been smashed open and my sister's dog had been stolen! Someone stole my sister's baby! Horrible!

The story made front page news in "The Vancouver Sun" on 7/29/2005. You can read the details there. Search on "Alert city police officer sniffs out two stolen dogs" in the on-line newspaper's search window.

Alert city police officer sniffs out two stolen dogs

Nicholas Read — Vancouver Sun — July 29, 2005
For Diane Anderson of Vancouver and the Langston family of
Everett, Wash., it was an answer to a month's prayers.
****

http://www.canada.com/search/results.html?searchfor=alert+city+p
olice+officer+sniffs+out+two+stolen+dogs
****

The point I wanted to make, without going into too much detail,
was the fact that I was on vacation in Yellowstone with my sister
before they found the dog. My sister had hired a pet psychic to try
and find her dog. The pet psychic phoned my sister when we were
in Yellowstone Park. She told her that she had a "vision" that my
sister's dog was stolen by someone of Asian decent and that it was
in good health. My sister also told me she had a dream the previous
night that revealed an Asian person had stolen her dog, as well.
The newspaper article brushes the psychic's facts off lightly.
Instead the article concentrates more on the fact that an alert police
officer found their dog. The officer had read a previously
published newspaper article about the lost dog. The police officer
phoned the breeder of my sister's dog to verify the tattoo numbers
they found on the dog.

I can see why police use psychics to break tough cases. If you were
a police officer that knew about certain people on your beat that
were being observed for possible thefts in the area. You then read
about a pet psychic that said a dog was stolen by a person of Asian
decent on your beat. It seems logical that the police officer could
become a little suspicious about the Asian person on their beat
being observed for possible thefts. That suspicion is how the police
officer found the dog. The officer simply knocked on the door of
the thief's residence and the stolen dog ran to the door to greet the
officer!

The Pighne Theory

Taking this idea further I think that we as a society are missing
something fundamental about nature's existence. I have tried to
label it as a "global consciousness". The idea is if something
intense has happened in the world that certain people are
passionate or frantic about then other "sensitive" souls can also
pick up those global "vibrations".

I also think there is more to how "life" functions than we know.
How do blood vessels know where to go to feed oxygen to the
cells of the body? How does sperm know to keep swimming until
it finds and egg to penetrate? How do salmon really find their way
home after swimming in the vast ocean for two to five years?
Practical solutions point to chemicals and other factors such as sun
position.

I think we need to stretch our minds a little and look further—
beyond our three dimensional world. We need to look at the
reasons why things really work in life by looking at how life
functions at a subatomic level— at the quantum physics level.
Physical rules, as we know them, do not exist at the quantum
physics level.

What if chemical structures need to be analyzed beyond our three
dimensional boundaries to really understand how they are used by
our cells? What if cells can work at the quantum level to find each
other without regard to space and time? What if each cell and
ultimately each person have a unique "cosmic" address that can be
used to find each other? Then it is easy to understand how blood
vessels find the cells that need the oxygen they carry...Maybe the
brain with all its specialized communication cells packed together
really can reach out and touch someone anywhere in our
universe— with less regard to space or time. If cells or chemicals

can use the capabilities found at the quantum physics level it makes what psychics say they are able to do just as normal as hearing a sound. Maybe that is why the same letters that spell psychics can be used to spell physics!

Maybe chemical bonds go beyond our world as we know it and function in the quantum world. I propose that cells can manipulate their needs in the quantum world. Cells can create chemical structures that function at a subatomic level and allow each cell to seek out their needs with less regard to space or time. This is just a logical extension of what we know about the universe. There is no reason to limit our model of the power and existence of God to only the three dimensional universe we live in.

I am proposing a theory I now will label the **Pighne Theory** (pronounced "pine"). The theory proposes that there is a part of life that exists beyond what we see— just like the roots of a pine tree. The roots of a pine tree are not visible but are very much a part of the pine tree and provide needed water and nutrients to the cells of the tree for survival. My **Pighne Theory** proposes that this root structure was created because the cells know ahead of time what they need to survive and they know where to get it. It is a conscious mass that exists beyond our three dimensional world that guides living things to find the resources they need to survive. This conscious mass may define God's existence in a fourth dimension. The essence of God in each of us has a link to this fourth dimension. Links to this conscious mass may be somehow stored in genes as part of the DNA structure. Cells are able to find the water and nutrients they need at a subatomic level with less regard to space or time because they have links to a conscious mass that guides them to the resources they need to survive with less regard to space and time. This theory stretches our minds beyond what we normally view but does not break any rules of physics. The quantum world is said to exist. All I am saying is that cells know how to use resources in this quantum world to find what they need to survive. Taking a closer look at the root structure we can see other similar looking structures in nature. If you rise above the earth and look down upon it one can come to a conclusion that a river and it's tributaries look like an upside down root structure with streams feeding creeks that turn into rivers that feed into other rivers and eventually into the ocean. We know that the river's root-like structure is, in large part, simply water flowing down hill and following the path of least resistance.

So why can't we apply the "least resistance" theory to the quantum world. Cells are simply finding what they need to survive at a subatomic level by following the path of "least resistance". The idea of "least resistance" follows soul searching techniques. To find your inner self you must relax, let go of the outer world, concentrate on a point within yourself and just let yourself go! Leaving the existing world to find peace of mind requires you to let all the worldly stress behind you and follow the path of "least resistance" within you. All kinds of structures in life look root-like including blood vessels and nerves. If you follow the roots of your past you end up discovering your "family tree". Our whole society is based on a hierarchical structure that can be interpreted as a root-like structure starting with the president of a company, for example. I think it is time we took this common structure and extended it to the subatomic level. I believe that all living cells have already figured out how to do that and utilize this root-like structure to link to the fourth dimension. Living things may find guidance from their ancestor's, from God, and other "resources" in what I am calling a "fourth dimension"!

The **Pighne Theory** puts a whole new twist on DNA, for example. I mentioned in my previous "Religion of One" article that the DNA structure may be nature's attempt to get "back to one" which is the fundamental driver for life's existence in us all. I am now proposing, in this article, that we all may have a cosmic address that is unique. This cosmic address allows us to find each other with less regard to space or time. What if genes are nature's way of storing that cosmic address? What if DNA is so special it is able to capture the location of what it needs to survive at the quantum level? The needs of cells are stored as addresses within each cell's genes. That is not too hard to swallow. Some would say that is already a known fact. But what if the chemicals that make up genes are guarded with such great care because they are able to tap into a dimension beyond our own and record the roots of their past and the cosmic requirements for their future in life! What if DNA is not just a code used to create life in the future but an address that can be used to find each unique entity in life, at a quantum level, in the past, present or future! Cells simply record what they need to survive in their genes. The cell's genes are a record of addresses of the locations of all the resources that are required for the cell to survive in the past, present and future! I propose that these addresses are supplied by following the path of "least resistance" at a quantum level to the addresses of the cosmic resources they need to survive.

Conclusion

I believe that Nature has created a recipe for survival inside genes that reach out to our cosmic universe at a quantum level using very special chemical structures that are fiercely protected in nature and are the holy grail of life. It is time we go a level lower than the atom to analyze how life really works. If cells can establish a need and fulfill that need by tapping into resources at a quantum level it opens up a whole new realm of possibilities! It opens up a whole new dimension! By proposing the Pighne Theory I am establishing a need to know more about how life interacts with our universe and beyond at a quantum physics level. In our quest to become one with God the journey to know how life interacts with our universe should bring us closer to God. If you follow my logic from the "Religion of One" article then God is the Universe!

There was and interesting article in Time's this week (08/08/2005). It covers some of the ideas I have presented here and calls it "Intelligent design". My input from some of the article's arguments about Darwin's theory and the 'Intelligent design" movement are as follows.

It seems to me that Darwin started in the middle of the evolution process by defining how species adapt. I think it is time that we go backwards from the middle and start looking at how cells adapt. I am calling my cell adaptation theory the "Pighne Theory". There is a lot more we know about how quantum physics may play an important role in defining the particles that make up the universe than in Darwin's time. At the quantum physics level some of the laws of nature as we know them do not exist. It is only a natural evolution of science to try and fit the quantum physics level into the whole model. God fits perfectly into the evolution model from the beginning of things. It is time we started to work back to the beginning from where we are now and open our minds to new ideas.

Fate

I have come up with the Pighne Theory that tries to explain paranormal behavior within a scientific context by suggesting that we must now consider how things behave at the subatomic level—at the quantum physics level. There is less regard for space and time at the quantum physics level.

It is my duty to record unbiased observations throughout my life that fit into this theoretical framework. One thing I have notice is that these observations are not a frequent occurrence.

I just gave a two week notice to my manager of my current job in a large company with over 50000 employees after almost eight years of service. I am going to take a job offer in a small company with a little over 100 employees. I took the job offer after experiencing the typical "put out to pasture" age discrimination scenario and frustration over growth opportunities in the company. Since I felt that I could still offer a lot to a company that needed my skills I was not ready to accept "the pasture"! I can say the precise moment of the decision to leave the company was emotionally charged. It was after a group meeting with my current manager. He told everyone that if they did not start filling out their time cards with the level of details that he required we would get an automatic below average rating on work performance for that year. I remember saying to myself. "I have got to get out of this place!" Who wants to work for a manager that bases a whole years performance on whether you fill out details on your time card correctly!!! About an hour later I got a phone call from a recruiter that I had been offered the job I had interviewed for three weeks earlier. Perfect timing! It must have been fate!

So where does the paranormal come in to play?

It is the observations I have made about specific dreams I have had.

There are, in fact, two dreams I have had in the past couple of years that come into play because I remember them clearly as not fitting into the reality of my existence at the time I had the dreams. Dreams are like that. They are a part of your subconscious. The subconscious works at a level where all practical rules are easily broken. Bizarre things can happen in dreams that do not fit reality as we consciously see it when we are awake. In dreams, bizarre behavior or situations are normal. Perhaps the subconscious reflects consciousness at the quantum physics level!

The first dream has me working in a company where there are clearly racks of computer equipment in a room— close in proximity to where my office desk is. That doesn't seem too out of the ordinary for a person responsible for maintaining large databases. The fact is the dream occurred about two years ago. It certainly did not fit into the context of the job I was in two years ago. The data center existed in another town several miles away. I was not near any racks of computer equipment.

I had no reason to record this dream publicly because it was just a dream and did not fit into the facts of my current situation. I just remembered the dream clearly and thought it was odd that I was dreaming about a job that appeared to show me in charge of computer hardware. I consciously knew I had no interest in a job taking care of computer hardware so I thought it was an odd dream. I actually found it a little disturbing because it clearly showed me in a job I know I would not like to do. I thought to myself, "Is this the future I have to look forward to? Yuck!"

This dream is about to become a reality. The small company I am going to work for has a raised floor room right next door to where my office desk is. I am not taking care of the computer hardware. I am taking care of the databases just like my current job. It makes sense now that my dream was showing me that I would be working for a small company where the racks of computers are located in a room that is in close proximity to my desk!

Some people would take this observation I have made to be fate. It was my fate to go work for this small company and my dream was just playing out the future reality of my existence two years before it happened!

I believe there is an emotional aspect to fate. It represents milestones in your life that are emotionally charged for one reason on another. They don't occur too often but represent major decision points in your life that shape the way you live after that milestone of fate has been reached. My second observation is the fact that we have a word in the English language that describes the events I have just experienced. I cannot say that I have ever witnessed an example of fate with such clarity in my life time. Fate clearly points to a point of need. I had a need to leave my current job. This need was fulfilled in my dreams two years before it happened!!!

I believe my second dream is also a milestone in my future life that has not occurred yet. It is a simple dream. Again my reaction at this point is that it is an odd dream, does not fit into the context of my current existence. I find the dream a little disturbing.

My dream is that I can see myself walking on a sidewalk in a very nice black suit. My physical appearance is with graying hair. I am smiling.

What I find odd about this dream is that I am walking alone. I have a family. Where are they? The other odd thing is that I am smiling.

My first assumption is that if I am wearing a black suit I must be coming from a funeral. Now that is not unusual but why am I smiling? I find that disturbing. There is something emotional going on there that, right now, I have no clue of. The fact is I know of several elderly people that are relatives and friends. I could see myself going to their funeral in the next few years.

It could also mean that it was my funeral!

In my dream, could I be walking away from my own funeral either knowing I had a good life or did something that left me in a happy state when I died? I plan to donate a kidney to my wife who has polycystic kidney disease in the next few years. Is it possible I did not fair so well but left her with a good kidney to survive on?

Who knows? I believe my future fate will reveal the facts behind this dream.

F=MA

I was helping my daughter with her science homework the other day. She was studying "$F = MA$" which is Newton's Second Law of Motion regarding the inertia of an object. What I realized is that paranormal forces can be described by this formula. An external energy called Need is what may generate paranormal forces. Paranormal forces can be described as any force that exists beyond our three dimensional world. Does conscious mass in the fourth dimension generate energy in the form of need that can transcend any dimension? This external energy affects the inertia of different types of conscious mass that exist beyond our own three dimensions.

Need (N) is a form of external energy required to get things moving. A Need can be defined as a conscious desire for something required by a living thing. Gasoline is energy required by a car to get it moving. Need is required by all living things to survive. Energy can be used to accelerate mass in a particular direction. Need can be used to accelerate a conscious mass towards fulfilling a requirement for survival.

Any Force is equal to Mass times Acceleration or $F = MA$. When you push on a ball you are using external energy to move the ball (mass) in a particular direction. At the subatomic level Acceleration is infinite so it becomes unimportant. All that is required to set a mass in motion in the fourth dimension may be the Mass itself and the Need to use it. This Need may be used by cells of living things to infinitely accelerate forms of paranormal, conscious matter in a direction that is required for survival! All living things may be able to communicate their needs into the fourth dimension and utilize the conscious mass that exists there. There may be many forms of the fourth dimensional, conscious mass directly dependent on the needs of living things! What that means is the manifestation of God could take many shapes depending on the needs of a group of living things. Maybe Christ came to earth and existed because a large mass of people on earth needed him. Maybe what I am calling fate is all my departed relatives watching out for me and guiding me from the other side! We have no idea of the scope of the fourth dimension. If it is anything like religious teachings tell us then possibilities and power of the resources that exist in the fourth dimension are endless. We should use them more often!

What I believe we do not understand yet is all the types of mass that exist at a quantum physics level and different dimensions. What it comes down to is that if you believe that there is a paranormal force then the strength of that force is directly related to the mass of the paranormal force and the Need that is applied to use that paranormal force. Need is used by all living things to Hunt (H) and Gather (G) the resources for survival. At a quantum level "H" may simply be a peak of a wave and the "G" may be the trough of a wave. When Need is applied to a conscious mass that exists beyond our three dimensional universe it becomes an awesome force that helps us survive. Living things are able to generate this need by converting the food they eat or the sun they harness into energy. This generation of needs occurs with less regard to space and time. Fat cells store energy for the body's future needs! How do they do it?

$E = MC2$ (C squared) according to Einstein.

There is a form of mass in the fourth dimension I am calling consciousness. It is a part of what was once One and guides us on our journey to become One again.

In the beginning there was just one mass and one force. The concept of Acceleration did not exist or it could be considered infinite. That one Force was God. That one Mass may be the conciousness of God as we know it in the fourth dimension. $F = MA$ where A is negligible. Therefore Force = Mass. God = Consciousness!

After the Big Bang the definition of mass became random and now comes in many more forms. Perhaps consciousness is simply a type of mass that all living things can use as a resource to help them survive. This conscious mass is able to exist beyond our three dimensional world.

Need generates the powerful instinct to hunt and gather for survival. What many religions teach us is that you have a conscious resource beyond this world can be used to help you survive. At the quantum physics level our cells and DNA may have already figured out how to do that.

I believe there are bonds at the quantum level we do not understand yet. These bonds may link us with a conscious mass of paranormal resources that exist in the fourth dimension. Bonds between random, wave-like structures may be a way of stabilizing what would otherwise be a totally random universe. The DNA structure is a window into how life, as we know it, interact with the wave-like structures at the quantum physics level. How these bonds work at a quantum physics level is still a great mystery.

Sleep allows us to form the bonds that are needed to survive using resources found in this world and beyond. Need to survive is based on a bonding of mass and light ($E = MC2$). Without those two things you are left with dark, empty space. What we see as dark space in the universe may manifest itself in the fourth dimension as a lack of consciousness. If consciousness is the state of being aware of your surroundings then lack of consciousness is the void we see between the stars in the universe. Religious teachings tell us that lack of awareness, random, chaotic behavior is the path to Hell. It makes perfect sense that lack of consciousness is not the right direction for living things to go if they want to survive. The very definition of consciousness means the state of being alive.

Paranormal forces have been simply described as a lot of "vibes" floating around. I believe it is more likely these "vibes" represent some sort of need. Need can be defined as the energy generated by living things to survive. Sensitive people (psychics) can pick up on these "vibes" and interpret them without even being consciously aware of it. These "vibes" generate inertia on a conscious mass that exists beyond our known universe. It could be argued that I am picking up those "vibes" and that is why I am writing this book! It is need that spawns paranormal forces. Need generates paranormal forces created by living things to survive. These forces can be explained as instincts or fate. Conscious desire can generate need. Maybe subconscious desire is used at a quantum level to help us survive. I believe consciousness is the essence of God. It is the fundamental building block of life. It is the desire in all living things to get back to One.

Need generates the hunt (H) and gather (G) instinct in all living things. The essence of God appears as a simple wave-like structure at the quantum level of physics in our universe. DNA is a wave-like structure. It follows that wave-like structures are the most basic structure in the universe. Maybe that is why we find waves so comforting! Babies love to hear the sound of a heart beat! A heart beat can be seen on a heart monitor as a simple wave pattern. Waves brings us closer to God.

The Pighne theory reaffirms what I believe is the infinite power of a wave. How wave-like structures interact beyond our three dimensional world is the key question I am asking our society to take a closer look at.

We have just started to understand the infinite power of waves.

A Twist of Fate

Thinking about the structure of DNA, further, I observed the following…

The structure of DNA is uniquely three dimensional. It has the shape of a simple wave. The wave twists into a three dimensional structure.

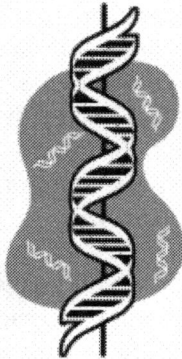

What does this really mean?

The DNA structure is reflecting our very existence in a three dimensional universe. This structure defines what it takes to survive in a three dimensional world.

What I am proposing, in my Pighne Theory, is the existence of a conscious "fourth dimension". Each strand of life is able to tap into this fourth dimension and utilize the resources that exist there. This is not anything new. It is the concept of Heaven and Hell! Why should there not be a fourth or fifth or sixth dimension? Just because the structure that makes up our human existence is only three dimensional does not mean there are not other dimensions beyond what we live in.

I am looking at the existence of a fourth dimension and stating that even in the scientific realm it makes sense!

I think a good experiment might be to weigh a fixed set of DNA strands in a Petri dish and then meticulously split apart each DNA strand in the Petri dish. After splitting the DNA strands apart I would weigh the total contents of the Petri dish again. It would not surprise me if the total weight of mass in the Petri dish was less than the total weight of the strands when the DNA was whole.

Why?

Perhaps splitting apart DNA releases the bonds to a conscious mass that exists in a fourth dimension. Would breaking the DNA bonds to this conscious mass decrease the total weight of the DNA structure? The concept of "afterlife" proposes a release of a conscious mass from our bodies after death. Maybe we can scientifically measure this fourth dimensional mass that is released. You cannot physically measure a fourth dimensional mass but maybe we can measure the amount of mass missing from the original DNA after destroying the bonds.

What I am proposing is that the DNA structure has a link to a fourth dimension. I believe there is a good possibility that there are forms of mass that we do not understand beyond our three dimensional world. Maybe paranormal consciousness is just form of "fourth dimensional" mass. Somehow, at the DNA level, all living things are able to capture this type of mass and utilize it for their need to survive.

Why is there a need to survive? What drives us to survive in this world? It is a force we call God. This life-giving force is made up of a conscious mass that exists at the lowest level of our known universe and beyond.

My theory proposes that Force may simply equal Mass at the quantum physics level because acceleration is infinite at that level. The greater the "fourth dimensional" conscious mass that exists then the greater potential "fourth dimensional" force there is. Need is what moves this conscious mass in the direction living things require it to move for survival.

Electromagnetic Field (EMF) meters are used to measure the strength of paranormal forces in a particular area. It would be an interesting experiment to go around and measure paranormal forces in areas where there have been a lot of deaths. For example, a battle field versus homes where there were only one or two deaths may be a good place to measure the volume of paranormal mass that exists in different areas of our world. Would the EMF reading be higher on a battle field than in the average home? A higher EMF reading on a battle field, versus an EMF reading taken in a home, makes sense if there is more paranormal, "fourth dimensional", conscious mass that exists on the battle field than in an average home. Perhaps electromagnetism is what DNA uses to bond the various resources it needs to survive as a conscious living thing.

What does this theory mean to every day living? It means that large masses of any kind can affect our very existence even if these masses exist in a dimension beyond the three that we exist in. All other outside forces being equal, the object with the greater mass will have greatest potential force to influence a specific situation. If you want to change the way you live or you want to make a difference in this three dimensional world we live in then you can use resources that exist in the "fourth dimension", as well as in our three dimensional world! Great forces can be generated from large masses. It is how the Roman army built an empire. Jesuit priests know that if they can teach children the beliefs of the Holy Catholic church at a young age then they have that child forever. By getting the idea of God into a child's subconscious at an early age the more likely that child will believe in those lessons for the rest of their lives. The priests are basically collecting a large mass of subconscious beliefs, child by child to produce a large religious force. Once you have a large mass that has the same subconscious beliefs that you do you can use the force of this mass to change the world! What most religions teach us is that there is a force beyond our three dimensional world that we can use to guide us through life. God exists. You can use God as a resource to help you fulfill your needs. I just glanced at a show on a public TV channel while randomly flipping channels to see what was on. The person was talking about the "Power of Intent". Looking on the internet there are a lot of articles about this. What I bring to this discussion is a scientific angle to it. The power of intent is simply the interaction of energy applied to a form of mass— the Need that is applied to Consciousness that exists in this world and beyond.

Sleep

Sleep is a wonderful thing. It filters out all the noise of the day and orders the mind's thoughts. It gives rest to the physical body and allows the body to store energy to deal with the needs of the next conscious day. Babies probably sleep a lot because their bodies are busy utilizing all the resources within our three dimensional world and beyond to ready their physical bodies for survival in a three dimensional world. Salesman are good at defining needs that never existed before. There are so many modern day needs that never existed before. It drives us to try and fulfill those needs to the point of exhaustion. Just talk to any mother in this world! I remember what a pastor told me about his conclusions of life based on all the issues he heard the people of his parish tell him about throughout the years. His conclusion was people were originally made to sit around, enjoy life, and eat bananas on a tropical beach. They were not made to deal with all the complex needs that modern society has thrust upon them.

After sleeping on the thoughts I have laid down in this text I have come to the following conclusions.

Need is a form of energy. Consciousness is a form of mass that goes beyond our three dimensional world into a fourth dimension and may be the essence of God. Consciousness is a state of being aware of one's surroundings. The idea that there is a consciousness beyond this three dimensional world is common place. Our three dimensional world is also simply a state of existence. Many religions define a state of consciousness or existence beyond our three dimensional world. A type of conscious mass that is in the "fourth dimension" explains paranormal forces that can be observed using EMF meters. Another observation is that when you come up with a new idea it is common to say "A light just turned on!" If you finally understand a concept that you were not previously aware of it is common to say "I finally see the light!" A light bulb is a universal symbol for an idea. An idea is a thought that fulfills a Need.

So Need is a form of energy, consciousness is a form of "fourth dimensional" mass and ideas equate to a light that can be described as something that gets things going in a direction that fulfills a need. What does that spell? $E = MC2$. Energy is equal to Mass times Light(C) squared! Need = a conscious Mass that sees the light(C) squared!

I could not quite remember what "C" stood for in this equation so I went and looked out on the internet for the definition. In the search results I found out that Einstein was a cosmic pantheist. He apparently believed in a "personal" God and felt something was missing in his definition of the universe.

My Pighne Theory suggests that there is nothing missing in his formula for the universe. What is missing is the recognition that consciousness is a form of mass that goes beyond our three dimensional world and is utilized as another resource for survival by all living things. It can be defined at the quantum level as just another piece of what was once One. It can be defined as the essence of God and what drives all conscious beings to survive. This conscious mass defines the very essence of our existence and is a part of the formula for our need to survive. It is what drives our need to become one with God again. The visualization of Christ on the cross may be a visual hint that if you die in our three dimensional world then it does not mean you cease to exist. If the cross represents our three dimensional world then Christ is showing us that you still exist after you die. You are just crossing over to a different state of consciousness!

Conclusion

The celebrations of various religions and holidays reflect on an existence that is beyond the three dimensional world we call home. These celebrations remind us that there is a conscious mass beyond our three dimensional world that we do not quite understand. We should try to observe these reflections in the light of modern physics and start measuring this "fourth dimensional" mass. I believe that the subconscious and paranormal forces are generated from various forms of "fourth dimensional" mass. This conscious mass is just another piece of the One we call God that existed before the "Big Bang".

It has been proposed that variations in oscillations of "strings" are what create the various elements in the universe. Different frequencies within each "string" define the great array of objects in the universe. I propose that these "strings" are the essence of God. A fundamental energy that drives our very existence is defined as the Need to be one with God. If there is a fundamental Need to be one with God then we can define that Need using Einstein's formula E=MC2. "E" represents the Need to be one with God. "M" represents a "fourth dimensional" type of mass I call consciousness and "C2" is simply the path of Light to God squared! The belief that in the afterlife you must follow a light to become one with God seems perfectly logical. This path of light is simply part of the formula that defines the Need to be one with God! Your "afterlife" conscious Mass must be combined with Light to fulfill this basic need. E = MC2 also defines the difference between Heaven and Hell. If you are not aware of your surroundings and mindful of the effects that this lack of consciousness has on other living things then you are going in the direction of darkness and the chaos of Hell!

We need to look at DNA beyond counting genes that are within each strand. We need to understand the total structure of DNA. We need to look at the bonds between the genes, the variations in the wavelike patterns that form the DNA structure and the links the chemical structures of DNA may have beyond our three dimensional universe. We are bound to God by the roots of our existence.

Marcas Major's

Post-Holiday Diet

Marcas Major

Introduction

Following my logic from the previous articles I realize now I have more power than I ever imagined. I can change my life for the better.

My efforts to lose or even maintain weight over the past 10 years have been dismal. My doctor told me it is normal for a mature adult to gain 2 to 4 pounds each year and never lose it. The reason is that your metabolism slows down and even if you eat the same amount you did 10 years ago you will probably gain weight. That being the case I have followed his advice to a tee. In high school I was around 175 pounds. Now at the age of 50, and right after the December holiday cheer, I was pushing 215 pounds with clothes on. Last year I was mostly at 208 pounds. From high school until now that is 40 more pounds I have to pack around. I am carrying around four sacks of potatoes every where I go! I think my weight should be about 185 pounds. I have never gone below 204 pounds in recent years.

A call for action came from my brother who is 58 years old. He told my mom that if I did not lose weight he was afraid I would not make it until I was 60 years old. My mom told me what he said and here I am writing this article. If anyone asks what I am doing I am going to tell them I am working on my PHD (Post-Holiday Diet)!

I have never tried to diet before but I am a logical guy that works out two or three times a week. I have a Bachelor of Science degree. I should be able to logically assess the facts and come up with a weight loss plan.

If you are not used to exercising, if you are extremely over weight, or if you have special issues to consider I would highly suggest you consult your doctor first. I am not a diet expert. I am just an average guy that wants to lose some weight after the holidays.

Dietary Facts I Have Learned Along the Way

1) Drink water first before grabbing a bite to eat. You might be just thirsty.
2) Obviously, the way I am eating right now has to change because I am gaining weight every year. I have to drastically change my food intake to lose weight.
3) Losing fat requires exercise. It is obvious that just trying to diet does not work. Personal trainers are popular with the rich folks and movie stars that have to keep their weight down. Dietary food products seem to have come a long way. I noticed friends on the latest fad diets have been losing a lot of weight! Most of them gain it back though.
4) You have to have the motivation to lose weight, write down everything you eat, and stick to a plan. Otherwise you are just doing what you always do and that obviously has not been a path to losing weight.

State of the Union

1) I have never been on a diet in my life.
2) I am not thrilled about running out and buying specialized low 'carb' food, day after day, and pretending I can keep this up for a very long time.
3) I am not thrilled about eating tuna fish straight from the can or throwing it on some raw lettuce for lunch every day.
4) I am overweight. I have learned that to work out any major problem you have to consciously solve it. You have to admit you have the problem and then you have to do something about it. If you keep the problem in the back of your mind (in your subconscious) and never directly deal with the problem then you are leaving it up to your subconscious to work it out for you. Your subconscious is what dreams are made of. There are no rules in the subconscious world— anything goes! That is why when people are subconsciously frustrated they eat too much. There is no logic in that behavior. These people are letting their subconscious take over their behavior. The subconscious comes up with some weird ideas on how to solve problems! You need conscious decisions to work out problems. Admit your problem and move on to solving it.
5) Protein shakes are used by many to lose weight quickly. They are a temporary solution. Permanent modifications in diet are required to maintain the weight lost by drinking protein shakes.
6) I exercise regularly so the problem is I eat too much!

Plan of Action

1) I will modify my diet to consume fewer calories by revising what I eat for breakfast and lunch.
2) I will continue to exercise.
3) I will maintain a log of what I eat and the weight lost for a two week period. I will not set a weight loss goal because I have never done this before.
4) I will drink water or decaffeinated mint tea before any meal.
5) I will use protein shakes to get my weight down because I don't want to fuss with special meal planning. I will eat a normal dinner without any restrictions except for NO SECOND HELPINGS!

After visiting my mom and having two chocolates and a chocolate éclair for dessert, I went to a grocery store knowing that I wanted to modify my breakfast and lunch calorie intake. I was looking for protein shakes. There was a big sale on a soy milk product. A rather large lady next to me took a couple of the soy milk containers and noticed I was looking at the same product after she selected it. She said that this product was good because it tasted ok and had no sugar in it. I took her advice and placed a couple of the soy milk containers in my cart. I then looked for a powdered protein shake product so I could just take it to my office and mix up a glass each day for lunch. I chose one that had the same artificial sweetener the soy milk had in it.

Date	Weight in pounds (Morning)	Breakfast	Snack	Lunch	Afternoon Snack	Dinner	After Dinner Snacks	Exercise
1/4/2005	211.5 (no clothes)	8 oz of water, 1 multi-vitamin and 8 0z soy milk	12 oz Mint Tea	2 scoops of Protein Shake Mix with 4 ice cubes	Gum and 12 oz Mint Tea;12 oz of water	10 oz of water; 1 glass of 2% milk; 1 Soft Taco with 2%	8 oz of water	Worked out for 1 hour. Ran and burned 390 Briskly
1/5/2005	210.5 (no clothes)	8 oz of water, 1 multi-vitamin and 8 0z soy milk	12 oz Mint Tea	2 scoops of Protein Shake Mix with 4 ice	Gum and 12 oz Mint Tea	milk;Lasa gna, Corn, Salad,5	8 0z soy milk (Chocolat e Flavored) 70	walked for 6 to 8 city blocks
1/6/2005	209.5 (no clothes)	8 oz of water, 1 multi-vitamin and 8 0z soy milk (Chocolat e Flavored)	12 oz Mint Tea	2 Pieces of Vegetaria n Pizza, 1 small piece of cake, 8 Oz 3 Oz Protein	Gum and 12 oz Mint Tea	10 oz of water; 4 oz soy milk (Chocolat e Flavored) 35 Calories;	4 Oz soy milk (Chocolat e Flavored) 35 Calories; 8 oz of water	Worked out for 1 hour. Ran and burned 390 Calories on tread mill for

Log 1

43

Date	Weight							
1/7/2005	208 (no clothes)	8 oz of water, 8 Oz soy milk (Chocolate Flavored) 70 Calories	24 oz Mint Tea	2 scoops of Protein Shake Mix with 4 ice cubes and water 140 calories	Gum and 12 oz Mint Tea	1 mulit-vitamin,10 oz 2% Milk,1 diet Rootbeer, Teriyaki Chicken meal (white meat only) small salad with two tablespoons of dressing; 4 cookies.	8 oz of water	Briskly walked for 6 city blocks during lunch.
1/8/2005	206.5 (no clothes)	8 oz of water, 1 multi-vitamin,8 Oz soy milk (Vanilla Flavored) 70 Calories	12 oz Water	2 scoops of Protein Shake Mix with 4 ice cubes and water 140 calories	12 oz Water	1 glass of red wine (6 Oz) ,Chile and Rice;Corn;4 cookies; 6 oz of 2% milk.	8 oz of water, 8 oz mint tea, 3 oz of soy milk (Vanilla Flavored)	Briskly walked for 6 city blocks during morning. Walkedwalked for 6 city blocks afternoon .

Log 2

Wow! That wasn't too hard. I lost five pounds in four days and only felt hungry a couple of times. I solved the hunger pangs by drinking a very small glass of soy milk.

Let's see what happens in the next few days!

Date	Weight							
1/9/2005	206.5 (no clothes)	8 oz of water, 1 multi-vitamin, 8 Oz Soymilk (Vanilla Flavored) 70 Calories	2 oz Water	2 scoops of Protein Shake Mix with 4 ice cubes and water 140 calories	1 small orange, 3 pieces of licorice, 1 2 oz Water	1 and a half glasses of 1% milk, 2 pieces of ham and pineapple pizza, 1 large cookie, 1 piece of licorice.		out for 1 hour. Ran and burned 390 Calories on tread mill for 20 minutes, 75 chin-ups at half my weight, 45 situps,
1/10/2005	206.5 (no clothes)	8 oz of water, 1 multi-vitamin, 8 Oz Soymilk (Capucci no Flavored) 70 Calories	12 oz Mint Tea	2 scoops of Protein Shake Mix with 4 ice cubes and water 140 calories	Gum and 12 oz Mint Tea	1 and a half glasses of 1% milk, teriyaki chicken and rice, 1 large coconut/c hoclate chip square, 3 pieces of licorice.	20 oz of water	Ran and burned 390 Calories on tread mill for 20 minutes
1/11/2005	206 (no clothes)	8 oz of water, 1 multi-vitamin, 8 Oz Soymilk (Capucci no Flavored) 70 Calories	12 oz Mint Tea	2 scoops of Protein Shake Mix with 4 ice cubes and water 140 calories	Gum and 12 oz Mint Tea	1 and a half glasses of 2% milk, Spinach filled Pasta, 1 coconut/c hoclate chip square, 4 pieces of licorice.	12 oz of Tea	Briskly walked for 6 city blocks at lunch.
1/12/2005	205 (no clothes)	8 oz of water, 1 multi-vitamin, 8 Oz Soymilk (Capucci no Flavored) 70 Calories	13 oz Mint Tea	2 scoops of Protein Shake Mix with 4 ice cubes and water 140 calories	12 oz Mint Tea	2 and a half glasses of 2% milk, Steak Stir Fry with Rice, 4 cookies, 4 pieces of licorice.	13 oz of Chamomile Tea	out for 1 hour. Ran and burned 390 Calories on tread mill for 20 minutes, 75 chin-ups at half my weight, 45 situps,

Log 3

45

Reality Check

Ok! Reality has set in. I reached a plateau. I will turn on the exercise a bit more and eat fewer cookies! People are telling me that what I am experiencing is loss of water weight. That is the easiest part of the diet. The hard part has just begun.

One observation, so far, is that alcohol should probably be cut out of any diet (maybe pizza and cookies too!). I had one glass of red wine and that is when I started to plateau. I think the wine goes straight to my waist. My guess is that those little yeast cells that made the wine did all the work my body should have done to get the grapes to a point where my body cells can suck it up and use it for energy. Alcohol is probably as close to fast food for the body you can get. My body probably just grabs that wine and, quick as can be, turns it into fat. I am starting to think that is how I got overweight in the first place. I never drank in high school. Two or three drinks a week when you are eating the same as you did ten years ago adds up to extra calories the body does not need! It is the extra few calories every week for 52 weeks a year that must make the two to four pound weight gain, each year, a reality.

Let's face it though, I am fifty years old and I am probably going to have a drink of wine a couple of days a week. It is something I enjoy. What I am starting to like about this diet is that it is so easy to do. I can see myself going on this diet every time I gain a few pounds. I always gain a few pounds around times of celebration during the year like birthday parties and holidays. The problem in the past is I never lost what I gained during those times of the year and I would just add to the gain on the next celebration or holiday.

I now have a tool I can pull out of the pantry to knock a few pounds off any time this happens! This diet has put me in total control of my weight for the first time in my life!

Date	Weight							
1/13/2005	204.5 (no clothes)	8 oz of water, 1 multi-vitamin,80z soy milk (Capuccino Flavored) 70 Calories	12 oz Mint Tea	2 scoops of Protein Shake Mix with 4 ice cubes and water 140 calories	1 roll of breath mints;12 oz Mint Tea	2 glasses of 2% milk, Spaghetti w/ French Bread, 4 cookies, 4 pieces of licorice.	12 oz of Chamomile Tea	Briskly walked for 6 city blocks at lunch.
1/14/2005	204.0 (no clothes)	8 oz of water, 1 multi-vitamin,80z soy milk (Capuccino Flavored) 70 Calories	12 oz Mint Tea	Chicken Burrito and a diet soda	2 rolls of breath mints;12 oz Mint Tea	2 scoops of Protein Shake Mix with 4 ice cubes and water 140 calories, 3 pieces of licorice	12 oz of Chamomile Tea; 4 oz of soy milk	No exercise.
1/15/2005	204.0 (no clothes)	8 oz of water, 1 multi-vitamin,40z soy milk (Chocolate Flavored) 35 Calories	12 oz Water	2 scoops of Protein Shake Mix with 4 ice cubes and water 140 calories	12 oz Mint Tea;4 pieces of licorice	Went out to dinner. 1 glass of wine before dinner;Mongolian Beef and Ginger Chicken with Broccoli; White rice; at a restaurant. Peach Ginger Tea;3 oz of Liqueur on ice	2 pieces of licorice;8 oz water	Ran 1.6 miles around neighborhood.

Log 4

Date	Weight								
1/16/2005	203.5 (no clothes)	8 oz of water, 1 multi-vitamin,8 0z soy milk (Chocolate Flavored) 70 Calories	12 oz Water	2 scoops of Protein Shake Mix with 4 ice cubes and water 140 calories	12 oz Hot Chocolate;1/2 pkg of a choclate covered, candy coated peanuts	1 glass of red wine (6 0z) ,Chicken Fajita;1 piece of lemon bundt cake; 8 oz of 2% milk.	12 oz of Mint tea	Ran 1.6 miles around neighborhood. Went snow skiing.
1/17/2005	202.5 (no clothes)	8 oz of water, 1 multi-vitamin,8 0z soy milk (Chocolate Flavored) 70 Calories	12 oz Mint Tea	2 scoops of Protein Shake Mix with 4 ice cubes and water 140 calories	24 oz Mint Tea	1 and a half glasses of 2%milk;hamburger quiche,salad with 1 tablespoon of dressing; 1 piece of lemon bundt cake	12 oz of Mint tea	No exercise.
1/18/2005	202.0 (no clothes)	8 oz of water, 1 multi-vitamin,8 0z soy milk (Chocolate Flavored) 70 Calories	24 oz Mint Tea	2 scoops of Protein Shake Mix with 4 ice cubes and water 140 calories	25 oz Mint Tea	All white Teriyaki chicken strips, rice, small salad with tablespoon of dressing; mint tea; two cookes; 3 pieces of licorice; 8 oz of 2% milk.	12 oz of Water;3 oz of soy milk .	Worked out for 1 hour. Ran and burned 390 Calories on tread mill for 20 minutes, 75 chin-ups at half my weight, 45 situps, 30 back situps.
1/19/2005	201.5 (no clothes)							

Log 5

Conclusion

Well, there you have it. I just lost about ten pounds in about fourteen days! That is one less sack of potatoes I have to carry around with me.

I went into this diet with no expectations of how much I would lose in two weeks because I had never done this before. My feelings at the start of this effort were that I could handle anything for just a couple of weeks.

At the end of the two weeks, I am surprised at the relative ease of this diet. There were few hunger pangs (I remember three of them in two weeks) and the hunger pangs were easily remedied by drinking an extra three or four ounces of soy milk. I have exercised regularly for several years so that was not a difficult thing for me to do.

This diet was not hard to follow. I ate a normal dinner almost every night. The diet was simple and flexible enough so that if I had a big lunch one day I would adjust my calorie intake by drinking a protein shake for dinner that night instead of the regular meal.

It is worth repeating that if you are not used to exercising, if you are extremely over weight, or if you have special issues to consider I would highly suggest you consult your doctor first. I am not a diet expert. I am just an average guy that wanted to lose some weight after the holidays.

I believe this diet is a wonderful way to lose weight because it is easily repeatable. It is a tool that you can use any time to knock down any excess weight you have. I now feel like I am in full control of how much I weigh!

Well, I take that back. I am not in full control of my weight. My wife was so amazed at the results of this diet she told me I need to keep going and lose more weight.

"You are doing so well!" she said.

As of this publication (six months later) I have maintained my weight at about 194 pounds dropping from a size 38 waist to a size 36 waist line.

About the Author

Marcas Major has a Bachelor of Science in Fisheries, a Master of Business Administration in Finance, and is a Microsoft Certified Database Administrator. Marcas Major is a pen name for Marcel Pighin. He lives with his wife and two daughters on Cougar Mountain, near Seattle, Washington.

www.ingramcontent.com/pod-product-compliance
Lightning Source LLC
Chambersburg PA
CBHW030304030426
42337CB00012B/582